SCOOBY-DOO EXPLORES TRANSPORT

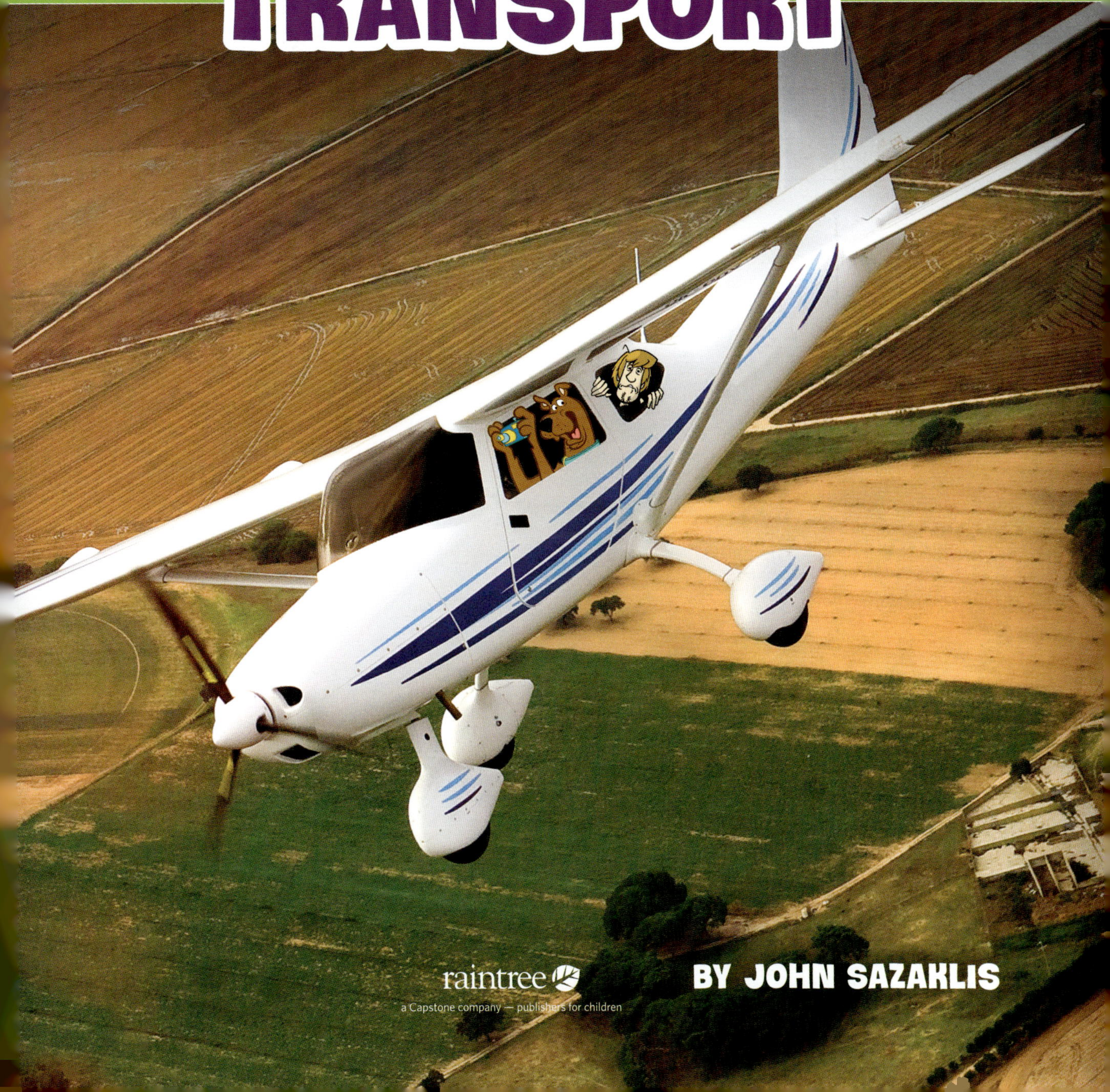

BY JOHN SAZAKLIS

Raintree is an imprint of Capstone Global Library Limited, a company incorporated in England and Wales having its registered office at 264 Banbury Road, Oxford, OX2 7DY – Registered company number: 6695582
www.raintree.co.uk
myorders@raintree.co.uk

Copyright © 2025 Hanna-Barbera.
SCOOBY-DOO and all related characters and elements © & ™ Hanna-Barbera. (s25)

The moral rights of the proprietor have been asserted. All rights reserved. No part of this publication may be reproduced in any form or by any means (including photocopying or storing it in any medium by electronic means and whether or not transiently or incidentally to some other use of this publication) without the written permission of the copyright owner, except in accordance with the provisions of the Copyright, Designs and Patents Act 1988 or under the terms of a licence issued by the Copyright Licensing Agency, 5th Floor, Shackleton House, 4 Battle Bridge Lane, London, SE1 2HX (www.cla.co.uk). Applications for the copyright owner's written permission should be addressed to the publisher.

ISBN 978 1 3982 5634 7

Editorial Credits
Editor: Christianne Jones; Designer: Bobbie Nuytten; Media Researcher: Rebekah Hubstenberger; Production Specialist: Whitney Schaefer

British Library Cataloguing in Publication Data
A full catalogue record for this book is available from the British Library.

Image Credits
Dreamstime: Mircea Mester, 17; Getty Images: Alexander Spatari, 16, andresr, 17 (middle right), artpartner-images, 31 (middle), clu, 7 (bottom), DAJ, 9 (middle), Drazen_, 4, EmirMemedovski, 20, Gary John Norman, 14, Hulton Archive, 7 (middle), 19 (middle), 27 (top), 31 (top), iStock/1933bkk, 21 (middle left), iStock/AntonMatveev, 5 (top), iStock/HaraldBiebel, 11 (top), iStock/imaginima, 28, iStock/manwolste, 21 (top right), iStock/MatusDuda, 29 (top right), iStock/Michael Hausmann, Cover (top right), iStock/RichieChan, 8, iStock/yanjf, 13 (middle right), magnez2, 17 (middle left), MarioGuti, 6, Rapeepong Puttakumwong, 19 (top), Rischgitz, 7 (top), Westend61, 12, 13 (middle left); Shutterstock: Aleksandar Malivuk, 23 (middle), cyo bo, 10, Denis Belitsky, Cover (middle right), Eroshka, 23 (top), fandangle, 23 (bottom), frank_peters, 31 (bottom), Grrrenadine, 11 (bottom), Guitar photographer, 11 (middle), Ira Gallo, 15 (top), Ivan Kurmyshov, 30, Leigh Trail, 15 (bottom), Ljupco Smokovski, 27 (middle), Luis Viegas, Cover (bottom right), 1, Matyas Rehak, 19 (bottom), Michele Ursi, 27 (bottom), NAN728, 15 (middle), Paul Vinten, 13 (top right), PeskyMonkey, 18, Skycolors, 29 (top left), Spiroview Inc, 25 (middle right), Standret, 25 (middle left), US 2015, 24, Venus Angel, 25 (top left), Vitaldesign, 5 (middle), Vitpho, 22, wavebreakmedia, 26, YesPhotographers, 9 (top right)

Printed and bound in India.

KEEP MOVING

The Mystery Machine has broken down, and Scooby-Doo and the gang have an important date at the annual Monster Convention. They'll need every type of transport to get there on time.

Use the clues in the text and the photos to guess which type of transport the friends are using to travel.

Scooby-Doo and his friends bump along in a big vehicle that makes a lot of stops. Some passengers get off at these stops. Others get on.

There are rows and rows of seats and windows.

It's not a fancy way to travel, but it follows a set route every day.

SCOOBY-DOO, WHERE ARE YOU?

The word *bus* is short for the Latin word *omnibus*, which means "for everyone".

The first bus line in the world opened in Paris, France, in 1662.

Carl Benz invented the automobile and the first motorised bus.

At the last bus stop, the gang hops off and heads to a busy station. They purchase tickets and wait and watch as people rush around.

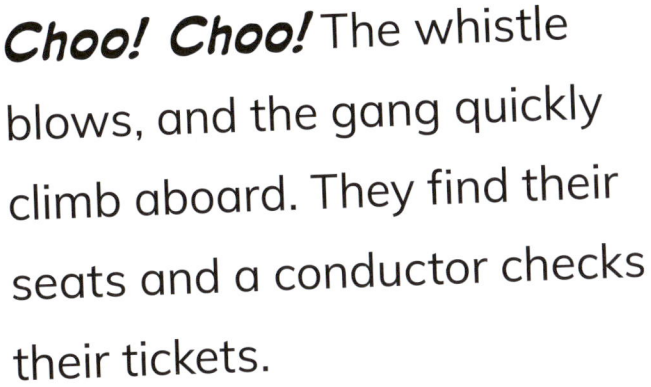

Choo! Choo! The whistle blows, and the gang quickly climb aboard. They find their seats and a conductor checks their tickets.

This vehicle chugs along rails. It can go short distances within cities or long distances across several countries.

SCOOBY-DOO, WHERE ARE YOU?

The first American railway was built in Lewiston, New York, in 1764.

Glacier Express

The fastest train in the world is the Shanghai Maglev train. The slowest is the Glacier Express in Switzerland.

The oldest working railway is Middleton Railway in Leeds, England. It was founded in 1758.

After the train ride, Fred decides the gang should split up. He ends up on a dock, looking out at sparkling water and rolling waves.

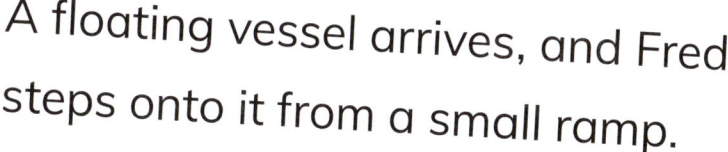

A floating vessel arrives, and Fred steps onto it from a small ramp.

Engines whir to life. The watercraft pushes forward, leaving ripples and bubbles in its wake.

The experienced driver in a special hat is called the captain.

FRED, WHERE ARE YOU?

FRED IS ON A BOAT!

"Sail away, Captain. I'm ready to "seas" the day!"

Ferryboats can carry people and cars across waterways.

There are many ways to travel by water, including cruise ships, fishing boats, kayaks, canoes, sailing boats, yachts and more!

A houseboat is a unique type of boat. It is a house and boat in one!

Daphne finds herself on a busy city street. She loves city life and knows a great way to travel.

Daphne raises her hand to hail a vehicle used for public transport.

It pulls over, and she climbs in the back seat. The driver asks Daphne where she wants to go. Now she can sit back and relax until her stop.

DAPHNE, WHERE ARE YOU?

DAPHNE IS IN A TAXI!

"What's worse than raining cats and dogs? Hailing taxis!"

People all around the world use the word *taxi*. It is a universal term.

The first taxis were horse-drawn carriages from England in the 1600s.

In Cuba, Coco taxis have three wheels. They are shaped like a coconut.

Honk! Honk! An old friend of Velma's pulls up in a huge vehicle with a loud horn. This very long mode of transport has 18 wheels. It carries things all over the country.

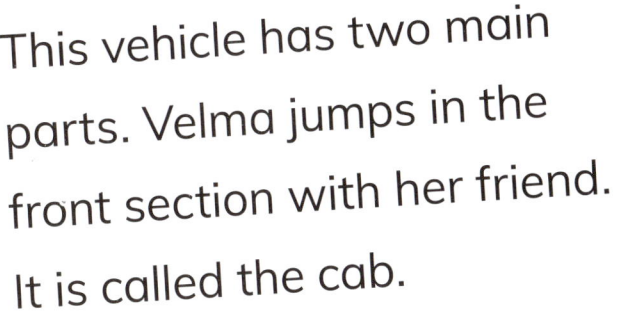

This vehicle has two main parts. Velma jumps in the front section with her friend. It is called the cab.

A trailer is attached to the cab. What is in the trailer? Could it be boxes and boxes of Scooby Snacks?

VELMA, WHERE ARE YOU?

VELMA IS IN A LORRY!

What scares a lorry? A monster truck!

The word *truck* comes from the Greek word *trochos*, which means "wheel".

Driving day and night is common for lorry drivers. Some lorries have a sleeper cab. The driver has a bed right in the lorry.

The longest trucking highway, Route 20, is 5,415 kilometres long. It stretches from Oregon to Massachusetts in the USA.

Shaggy and his furry friend are on the hunt for a simple form of transport. They see exactly what they are looking for lined up on a rack on the pavement.

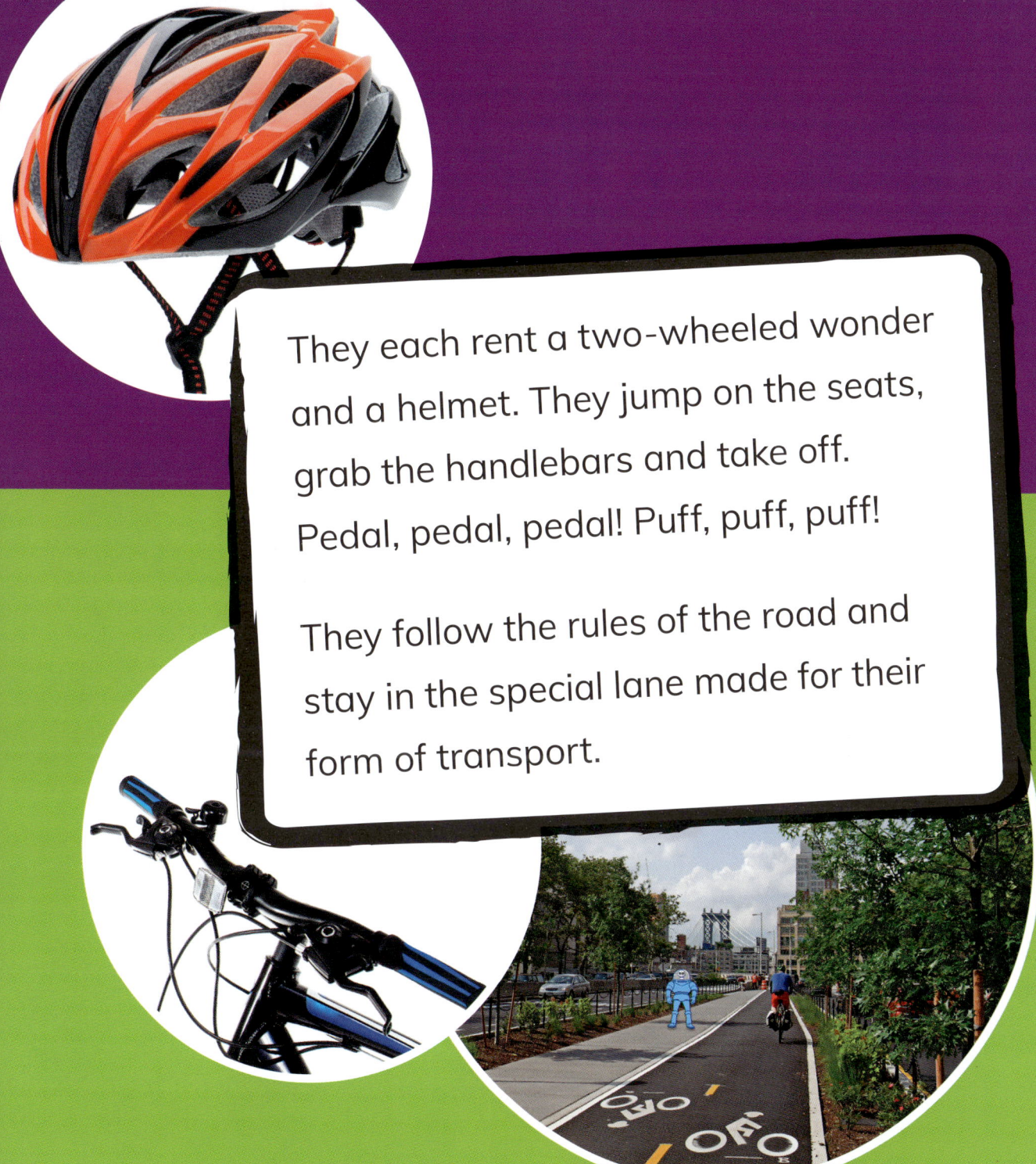

They each rent a two-wheeled wonder and a helmet. They jump on the seats, grab the handlebars and take off. Pedal, pedal, pedal! Puff, puff, puff!

They follow the rules of the road and stay in the special lane made for their form of transport.

SCOOBY-DOO AND SHAGGY, WHERE ARE YOU?

SCOOBY-DOO AND SHAGGY ARE ON BIKES!

"Like, safety first, Scoob. That's how we roll!"

"Rhat's right!"

In 1817, a German man named Karl von Drais created the first bicycle. It had two wheels but no pedals.

A tandem bicycle is made for two people and has two seats and two sets of pedals.

There are more than 1 billion bicycles in the world. Denmark has more bikes than cars.

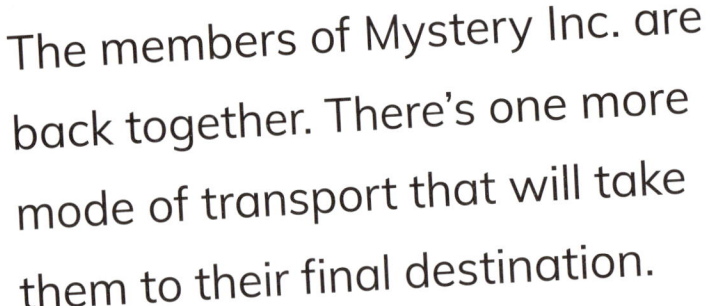

The members of Mystery Inc. are back together. There's one more mode of transport that will take them to their final destination.

This large flying machine has room for hundreds of passengers. It has high-speed engines and wings on the side.

The drivers are called pilots. They sit in a control room called the cockpit.

The gang buckles their seatbelts, ready for take off!

SCOOBY-DOO, WHERE ARE YOU?

The Wright brothers built the first powered aeroplane. Its first flight was on 17 December 1903.

Planes are often struck by lightning, and that's okay. They are designed to be struck without causing damage.

Most commercial aeroplanes have two large engines, but they can safely fly with one.

Scooby-Doo and the Mystery Inc. gang made it to Monster Con! They used many modes of transport, but they weren't alone. Did you notice any villains catching a ride with the group? Look through the book again and see who you can find!

ABOUT THE AUTHOR

John Sazaklis is a *New York Times* bestselling author with almost 100 children's books under his utility belt! He has also illustrated Spider-Man books, created toys for MAD magazine and written for the BEN 10 animated series. John lives in New York City, USA, with his super-powered wife and daughter.